Simulating Chinese Gray Zone Coercion of Taiwan

Identifying Redlines and Escalation Pathways

RAYMOND KUO, CHRISTIAN CURRIDEN, CORTEZ A. COOPER III,
JOAN CHANG, JACKSON SMITH, IVANA KE

Prepared for the Defense Threat Reduction Agency
Approved for public release; distribution unlimited

NATIONAL DEFENSE RESEARCH INSTITUTE

For more information on this publication, visit **www.rand.org/t/CFA2065-1**.

About RAND

The RAND Corporation is a research organization that develops solutions to public policy challenges to help make communities throughout the world safer and more secure, healthier and more prosperous. RAND is nonprofit, nonpartisan, and committed to the public interest. To learn more about RAND, visit www.rand.org.

Research Integrity

Our mission to help improve policy and decisionmaking through research and analysis is enabled through our core values of quality and objectivity and our unwavering commitment to the highest level of integrity and ethical behavior. To help ensure our research and analysis are rigorous, objective, and nonpartisan, we subject our research publications to a robust and exacting quality-assurance process; avoid both the appearance and reality of financial and other conflicts of interest through staff training, project screening, and a policy of mandatory disclosure; and pursue transparency in our research engagements through our commitment to the open publication of our research findings and recommendations, disclosure of the source of funding of published research, and policies to ensure intellectual independence. For more information, visit www.rand.org/about/principles.

RAND's publications do not necessarily reflect the opinions of its research clients and sponsors.

Published by the RAND Corporation, Santa Monica, Calif.
© 2023 RAND Corporation
RAND® is a registered trademark.

Library of Congress Cataloging-in-Publication Data is available for this publication.
ISBN: 978-1-9774-1122-8

Cover image: *Tuna salmon/AdobeStock*

About These Conference Proceedings

Taiwan has suffered extensive and escalating gray zone coercion from Beijing. China uses a combination of political, military, economic, and diplomatic tools short of war to put pressure on Taipei; restrict its international space and sovereign recognition; highlight the consequences of failing to accommodate Beijing; and normalize Chinese control of Taiwan's maritime and air space. These activities also aim to deter outside intervention in cross-Strait activity by the United States and its allies. But is that deterrence effective? Or will it escalate to conventional military conflict between China on one side and Taiwan and possibly the United States on the other side? Which scenarios are more escalatory, and where are the U.S. and Taiwanese redlines for war? To address these questions, the RAND Corporation conducted a series of discussions with security experts in Taiwan and with U.S. Indo-Pacific Command personnel and hosted a tabletop exercise in November 2022 in Washington, D.C. This paper details the key findings.

The conference proceedings here were completed in January 2023 and underwent security review with the sponsor before public release.

RAND National Security Research Division

This research was sponsored by the Defense Threat Reduction Agency (DTRA) and conducted within the International Security and Defense Policy Center of the RAND National Security Research Division (NSRD), which operates the National Defense Research Institute (NDRI), a federally funded research and development center sponsored by the Office of the Secretary of Defense, the Joint Staff, the Unified Combatant Commands, the Navy, the Marine Corps, the defense agencies, and the defense intelligence enterprise.

For more information on the RAND International Security and Defense Policy Center, see www.rand.org/nsrd/isdp or contact the director (contact information is provided on the webpage).

Acknowledgments

The authors of this paper are grateful to DTRA for its sponsorship of this work, particularly for the guidance and support provided by Robert Peters and Michael Sims at DTRA. We also thank our reviewers, Mark Cozad and Andrew Scobell, for their thoughtful and insightful reviews.

Summary

Taiwan has suffered extensive and escalating gray zone coercion from Beijing. China uses a combination of political, military, economic, and diplomatic tools short of war to put pressure on Taipei; restrict its international space and sovereign recognition; highlight the consequences of failing to accommodate Beijing; and normalize Chinese control of Taiwan's maritime and air space. These activities also aim to deter outside intervention in cross-Strait activity by the United States and its allies. But is that deterrence effective? Or will it escalate to conventional military conflict between China on one side and Taiwan and possibly the United States on the other side? Which scenarios are more escalatory, and where are the United States and Taiwanese redlines for war? To address these questions, the RAND Corporation conducted a series of discussions with security experts in Taiwan and with U.S. Indo-Pacific Command personnel and hosted a tabletop exercise (TTX) in Washington, D.C. This paper details the following key findings:

- The risk of escalation from the gray zone to conventional war is low because of Chinese military power and Chinese and U.S. constraints on Taiwan's decisionmaking.
- The TTX's U.S. team would only consider escalation to military force if Taiwan faced an existential threat, such as an invasion or blockade.
- The TTX's Taiwan team and Taiwanese experts would not escalate without U.S. support.
- The TTX's U.S. team was significantly more willing to intervene and provided more support when Taiwan took steps to defend itself in line with U.S. strategy.

In addition, RAND identified several broader factors inhibiting closer U.S.-Taiwan security cooperation. These include the following:

- The United States and Taiwan are caught in a prisoner's dilemma. Their security strategies hinge on one another, but each side is waiting for the other to signal deeper commitment before they will implement collective policies further.
- Some TTX participants worried about providing Taiwan with a blank check that would encourage adventurism. By contrast, Taiwanese experts and the TTX's Taiwan team refused to act without U.S. support, if not authorization.

Contents

Figures and Tables

Figures

Tables

Chapter 1. Introduction: China's Gray Zone Challenge to Taiwan

Beijing has long used a variety of gray zone tactics—acts of aggression short of war—against Taiwan. It has interfered in the island's politics, menaced its airspace, surrounded it with military drills, and sought to reduce Taipei's international space.[1] In doing so, China has sought not only to coerce Taiwan but also to warn off any intervention from the United States or Japan.[2]

To address these issues, the RAND Corporation conducted a series of discussions with security experts in Taiwan and with U.S. Indo-Pacific Command (INDOPACOM) personnel and hosted a tabletop exercise (TTX) in Washington, D.C. Our research suggests that the risk of escalation from gray zone aggression to war may be relatively low. The United States seems unlikely to allow itself to become militarily involved unless the viability of the Taipei regime is threatened. Most actions in the gray zone would leave the Taipei government intact and in control of the island, however much these actions threaten its control over surrounding waters. While the Taiwanese government resents these encroachments, it might be reluctant to act without clear U.S. support. Greater Taiwanese efforts to implement the overall defense concept (ODC) and build asymmetric, survivable forces could lead to greater U.S. support, although Taipei may prove unwilling to take such steps and reduce its investments in conventional forces before it receives clearer indications of U.S. backing. We caveat these findings by noting that these were the views of the experts we interviewed and who participated in our TTX, and more research is needed to determine whether these views are applicable outside a Taiwan crisis.

This chapter begins by establishing our definition of *gray zone coercion.* The subsequent section briefly details these Chinese efforts against Taiwan. The final section will outline the rest of this paper.

Defining the Gray Zone

This paper follows Lanoszka in defining the *gray zone* as a strategy of political coercion. Practitioners use and combine military, paramilitary, unconventional, or irregular capabilities to force a target to comply with a set of preferred policies.[3]

---

[1] Chin-Kuei Tsui, "China's Gray Zone Activities and Taiwan's Response," Stimson Policy Memo, Henry L. Stimson Center, December 12, 2022.

[2] Benjamin Jensen, Bonny Lin, and Carolina G. Ramos, "Shadow Risk: What Crisis Simulations Reveal About the Dangers of Deferring U.S. Responses to China's Gray Zone Campaign Against Taiwan," Center for Strategic and International Studies, February 16, 2022.

[3] Alexander Lanoszka, "Russian Hybrid Warfare and Extended Deterrence in Eastern Europe," *International Affairs*, Vol. 92, No. 1, January 2016.

Importantly, coercers must possess *escalation dominance*: the ability to "engage and defeat its target at different levels of military escalation."[4] This includes conventional war but can also encompass paramilitary (e.g., militias), unconventional (e.g., cyber warfare, nuclear threats), or even civilian domains (e.g., coast guards). Belligerents select a low rung on the escalation ladder, paired to a policy demand. Targets are then forced either to concede or to escalate, where they will again be at a disadvantage and might suffer greater costs.

However, coercers often lack global escalation dominance. More-direct use of force might elicit resistance from a militarily superior coalition of adversaries. If the target has powerful allies or friends, hybrid warfare also helps avoid triggering an intervention that the belligerent does not believe it can handle.[5]

Consequently, gray zone strategies direct sufficient coercive power against local, individually weaker targets to secure compliance with the belligerent's policies. However, those efforts are simultaneously limited or designed to avoid retaliation by stronger, often extraregional states.

While gray zone strategies are not new, they have attracted growing attention in the past decade as both China and Russia have increasingly turned to these methods to aggressively pursue national objectives without risking wider military conflict.[6] Recent work has included reports analyzing deterrence of gray zone actions,[7] economic coercion and trade quarantines,[8] and a close examination of gray zone actions in the South China Sea.[9] We hope this paper can contribute to the body of gray zone literature by examining some of the ways in which such actions could escalate into an unintended war within the context of China's conflict with Taiwan.

In choosing to focus on gray zone aggression, we deliberately chose not to examine the implications of a full-scale Chinese invasion or full military blockade of Taiwan. While questions about how such a conflict would likely proceed, the chances that the Chinese military could conquer the island, how it could be thwarted, and the impact of a great-power war between the United States and China are all important topics worthy of thorough research, they have already been extensively explored. Many games, models and simulations, and analyses covering

---

[4] Lanoszka, 2016, p. 178.

[5] Lanoszka, 2016, p. 180.

[6] James Goldrick, *Grey Zone Operations and the Maritime Domain*, Australia Strategic Policy Institute, October 2018; Geraint Hughes, "War in the Grey Zone: Historical Reflections and Contemporary Implications," *Survival*, Vol. 62, No. 3, May 14, 2020.

[7] Michael J. Mazarr, Joe Cheravitch, Jeffrey W. Hornung, and Stephanie Pezard, *What Deters and Why: Applying a Framework to Assess Deterrence of Gray Zone Aggression*, RAND Corporation, RR-3142-A, 2021.

[8] Bradley Martin, Kristen Gunness, Paul DeLuca, and Melissa Shostak, *Implications of a Coercive Quarantine of Taiwan by the People's Republic of China*, RAND Corporation, RR-A1279-1, 2022.

[9] Center for Strategic and International Studies, "Asia Maritime Transparency Initiative," webpage, undated.

a full-scale war over Taiwan have already been conducted. [10] We hope our paper can shed some light on the complex relationships between gray zone action and full-scale military conflict, particularly around Taiwan.

Chinese Gray Zone Coercion Against Taiwan: Goals and Efforts

China has targeted Taiwan with significantly more coercive attempts than it has other Asian countries and for a longer period (see Figure 1.1).[11] Moreover, these challenges have escalated significantly in the past several years. A record number of aircraft (196) intruded on Taiwan's air defense identification zone (ADIZ) ahead of Xi Jinping's speech in October 2021 (see Figure 1.2).[12] More than double that number (446) intruded in August 2022 following U.S. House Speaker Nancy Pelosi's visit to Taipei, with some crossing the Taiwan Strait's median line (the unofficial border between the two countries) for the first time. Beijing also bracketed the main island with military exercises and missile launches and has generally maintained a heightened tempo of incursions since. China has escalated its cyberattacks and disinformation campaign against Taiwan as well.[13]

---

[10] Many of games are conducted by the U.S. military and are classified. For a recent unclassified, third-party game analyzing this scenario, see Mark F. Cancian, Matthew Cancian, and Eric Heginbotham, The First Battle of the Next War, Center for Strategic and International Studies, January 2023. Many models and simulations are built by or under the direction of the U.S. military and are not publicly available. For one set of openly available model- and simulation-based analyses, see Eric Heginbotham, Michael Nixon, Forrest E. Morgan, Jacob L. Heim, Jeff Hagen, Sheng Li, Jeffrey Engstrom, Martin C. Libicki, Paul DeLuca, David A. Shlapak, David R. Frelinger, Burgess Laird, Kyle Brady, and Lyle J. Morris, The U.S.-China Military Scorecard: Forces, Geography, and the Evolving Balance of Power, 1996–2017, RAND Corporation, RR-392-AF, 2015. The analyses include Joel Wuthnow, Derek Grossman, Phillip C. Saunders, Andrew Scobell, and Andrew N. D. Yang, eds., Crossing the Strait, National Defense University Press, 2022; Timothy R. Heath, Kristen Gunness, and Tristan Finazzo, The Return of Great Power War: Scenarios of Systemic Conflict Between the United States and China, RAND Corporation, RR-A830-1, 2022.

[11] Bonny Lin, Cristina L. Garafola, Bruce McClintock, Jonah Blank, Jeffrey W. Hornung, Karen Schwindt, Jennifer D. P. Moroney, Paul Orner, Dennis Borrman, Sarah W. Denton, and Jason Chambers, Competition in the Gray Zone: Countering China's Coercion Against U.S. Allies and Partners in the Indo-Pacific, RAND Corporation, RR-A594-1, 2022b.

[12] Brian E. Campbell, "Record-Setting Incursions into Taiwan's Air Defense Identification Zone: The People's Republic of China's Psychological Operations Designed to Erode U.S. Support for Taiwan," Journal of Indo-Pacific Affairs, August 1, 2022.

[13] Yamaguchi Shinji, Yatsuzuka Masaaki, and Momma Rira, China's Quest for Control of the Cognitive Domain and Gray Zone Situations, National Institute for Defense Studies, January 2023; Yimou Lee, "Taiwan Defence Ministry: Website Hit by Cyber Attacks amid China Tensions," Reuters, August 3, 2022.

Figure 1.1. Chinese Coercive Activity Against Select Countries

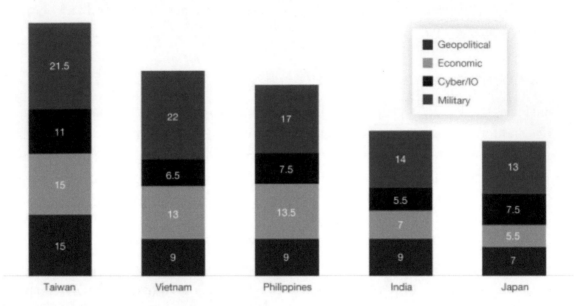

SOURCE: Reproduced from Lin, Garafola, et al., 2022, p. vi.
NOTE: Counts of People's Republic of China (PRC) tactics may not sum to whole numbers because of the way each tactic was coded. A 1 indicates relative confidence that China has used the tactic, while 0.5 indicates a suspicion that China has used the tactic against an ally or partner. IO = information operations.

Figure 1.2. Chinese Incursions into Taiwan ADIZ

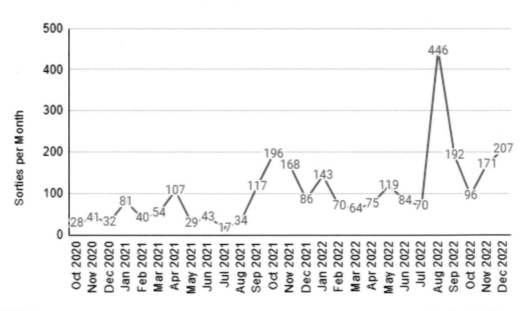

SOURCE: Brown, Gerald C., and Ben Lewis, "Taiwan ADIZ Violations," spreadsheet, March 5, 2023.

Research Questions and Outline

For gray zone strategies to work, coercers assume that their targets will fear a military loss and avoid the costs of military engagement. Belligerents can thereby achieve their policy goals while keeping conflict at a relatively low level and avoiding outside intervention.

But only the target knows its own threshold for military escalation. Similarly, only third parties know their thresholds for intervening militarily. Identifying these thresholds is therefore necessary to avoid misperception and miscalculation. Regarding China and Taiwan, we asked three major research questions:

1. What Chinese gray zone scenarios are most likely to escalate to conventional war, especially one involving the United States?
2. What factors most contribute to escalation, and how can those be mitigated?
3. What are the gaps in U.S. and Taiwanese preparations and responses to Chinese gray zone challenges?

Chapter 2 will detail our methodology. In brief, we followed a three-step research process. First, we surveyed RAND experts on the most likely and most escalatory Chinese gray zone challenges to Taiwan. Second, the team presented the top scenarios to security experts in Taiwan and at INDOPACOM to elucidate their expectations about how the United States, Taiwan, and China would respond. Third, we convened a TTX to determine how these responses would shift as each party interacted and affected escalation risk.

Chapter 3 details the main takeaways from the TTX. Moreover, several additional policy implications emerged from the research process. These are delineated in Chapter 4, which also highlights further policy questions for future research.

Chapter 2. Gray Zone Scenarios

To explore the ways in which gray zone aggression can escalate into full scale war, we devised a list of hypothetical scenarios in which the PRC commits various acts of aggression against Taiwan. In choosing these scenarios, we sought for acts of aggression that were plausible using past Chinese actions and that posed a particularly great danger of escalation. This included some actions that we recognize are improbable but that we believe are likely enough and dangerous enough to merit consideration. While the probability that China attempts any one of these actions in the future might be low, we believe that as tensions continue to mount across the Taiwan Strait, China is likely to continue to escalate its gray zone aggression against Taipei.

Scenario Selection

To cover as broad an array of plausible future scenarios as possible, we first generated a set of 13 possible Chinese acts of aggression short of full-scale war and four scenarios instigated by Taiwan or the United States. These scenarios were built largely using past Chinese and Taiwanese actions, although most included escalation to levels of aggression not yet displayed by either. Some included especially escalatory gray zone actions that would be unprecedented. Others were more conservative. Some, such as missile overflights and drone operations, occurred during the research. While we recognized that some of the scenarios we selected (such as political assassination or nuclear tests) might be highly unlikely, we tried to cover as broad a variety of acts of aggression as possible. Even improbable scenarios may merit consideration if they are both plausible and highly dangerous. A complete list of these scenarios is found in Appendix A.

Scenario Assessment by RAND Subject-Matter Experts

After generating our initial list of scenarios, we asked RAND subject-matter experts (SMEs) to assess the likelihood of China taking such actions and the potential for escalation to a full-scale conflict. The experts were asked to rate each scenario on a scale of low to high likelihood (i.e., whether the scenario was likely to occur) and, separately, low to high escalation potential (i.e., whether it was likely to lead to conventional war). The results of this examination are summarized in Table 2.1.

Table 2.1. Gray Zone Scenario Likelihood of Occurrence and Escalatory Potential

Likelihood	Escalation Potential		
	Low	Medium	High
High	Taiwan military Interference Hostage-taking		
Medium	Economic coercion Weapons interdiction	Missile/unmanned aerial vehicle (UAV) overflight Single port blockade	Blockade
Low	Cyberattack Nuclear test Refugee flow	Manned overflight	Island seizure Assassination

SOURCE: Data from RAND SME surveys.

As the table indicates, some scenarios were believed to be highly likely and some were believed to be highly escalatory, but never both. The closest was a Chinese blockade or quarantine (in which China declares its right to regulate Taiwanese trade and retaliates against companies that continue to trade or transport goods to or from the island without approval), which was believed to be moderately likely and highly escalatory.

We also asked RAND researchers to rate the likelihood of several scenarios in which Taiwan took actions short of war that could escalate to a large-scale conventional conflict. These were developed using past Taiwanese actions and the possible mirroring of some Chinese gray zone actions. The scenarios included an invitation to the United States to base combat troops on the island, highly nationalist and explicitly separatist educational programs, a reboot of the Taiwanese nuclear weapon program, and cyberattacks on PRC infrastructure. We felt that these scenarios were either unlikely to occur, unlikely to lead to escalation, or both, so we did not pursue them further. As will be discussed in Chapter 4, the results of our TTX further reinforced this conclusion, with the Taiwan (Green) team careful not to provoke China without U.S. backing despite a desire to push back strongly on new Chinese acts of aggression.

Scenarios Used in Interviews with External Experts and TTX

Using input from RAND SMEs and sponsor discussions with the Defense Threat Reduction Agency (DTRA), we selected and further developed four scenarios. We then approached SMEs in Taiwan and INDOPACOM for their assessment of how and how likely each scenario would lead to large-scale conflict between Taiwan, China, and the United States. The SMEs included academics, scholars at defense-oriented think tanks, former and current government officials, and military officers. The four scenarios were as follows:

1. *Maritime Interdiction*: A China Coast Guard (CCG) ship approaches a Taiwanese civilian freighter carrying electronics and machinery transiting through the South China Sea en route to Taiwan. The CCG vessel announces that its crew will board the freighter and search for contraband, because the ship is allegedly operating within "historical Chinese waters" without permission. The Taiwanese vessel refuses, and the CCG ship rams it. The freighter takes on water. The crew evacuates and is rescued by the CCG ship.
2. *Redirection of Maritime Shipping*: The PRC declares that it is exercising its right as the sovereign government of Taiwan to regulate the island's trade and that ships must first stop in a PRC port to clear customs and inspection before heading to Taiwan. Violators will be barred from shipping any goods to or from the Chinese market; their assets in the PRC may be subject to freeze or seizure to pay fines; and their vessels could be subject to impoundment or even attack.
3. *Interference with Taiwan Military Operation*: A Taiwan Coast Guard (TCG) vessel en route to resupply Itu Aba is surrounded by about 200 Chinese fishing boats. Behind them is a CCG cutter and several destroyers cruising within missile range. The Taiwanese ship is forced to ram its way out, sinking a Chinese fishing boat. PLA Navy ships then launch a heliborne assault and successfully seize the Taiwanese ship. No one is killed; there are minor injuries among the Taiwanese crew.
4. *Chinese Military Tests*: The PLA flies a dozen military drones over the Diaoyu/Senkaku Islands in a simulated attack on dummy maritime targets east of the islands.

Following interviews with U.S. and Taiwanese experts, we made slight modifications to these scenarios before settling on a final set of scenarios and phases for the TTX. For example, several experts believed that provocations far from the island of Taiwan would be unlikely to lead to further escalation, so some scenarios were modified to bring them closer geographically. In addition, China began overflying outlying Taiwanese islands with drones during the project, validating that scenario. It was updated to include deeper incursions.

Additional insights drawn from our discussions with these SMEs can be found in Chapter 4 of this paper.

Scenarios for the TTX and TTX Design

To further examine possible escalation dynamics in a gray zone confrontation, we conducted a TTX with several former U.S. civilian and military officials, defense policy experts, and experts in Taiwanese policy and politics. Recently, international relations scholars have increasingly turned to TTXs and wargames as analytical and research tools.[14] These methods are more immersive than other expert elicitation approaches and can help model the interaction and decisionmaking among actors. This elicits responses that reflect and respond to real-world

---

[14] See for instance Erik Lin-Greenberg, Reid Pauly, and Jacquelyn Schneider, "Wargaming for International Relations Research," *European Journal of International Relations*, Vol. 28, No. 1, 2021; Erik Lin-Greenberg, "Wargame of Drones: Remotely Piloted Aircraft and Crisis Escalation," *Journal of Conflict Resolution*, Vol. 66, No. 10, June 6, 2022; Peter Perla and E. D. McGrady, "Why Wargaming Works," *Naval War College Review*, Vol. 64, No. 3, Summer 2011.

political, military, diplomatic, economic, and social incentives, useful for predicting and evaluating actual state decisions.

However, the results from wargames fundamentally depend on their design, structure, rules, adjudication, and players. Researchers must carefully balance

- realism: to accurately model the incentives that actual decisionmakers face, tempered with the risk of overwhelming detail and constraints
- choice: to ensure that players can make meaningful and reflective decisions
- analytical utility: to force attention to the specific decisions the research is meant to assess
- participants: to ensure that players—because of their backgrounds, experience, and analytical processes—mirror those of actual decisionmakers.

To clarify, this chapter will refer to the TTX teams as Blue (the United States), Green (Taiwan), Red (China), and White (the control cell). Moreover, unless otherwise specified, the decisions, motivations, and concerns presented here reflect only those of the TTX participants. The RAND investigators themselves served as the Red and White teams. Both the Blue and Green teams were presented with the first part of each scenario and moved to different rooms. We then asked them to discuss their preferred courses of action, why they favored these courses of action, what each hoped the other team would and would not do, and their concerns about potential escalation. Both teams were allowed to send a few delegates to a joint discussion in the middle of the play session, after which they returned to their rooms. After an hour and a half, both teams joined a plenary session in which both discussed their preferred courses of action, differences of opinion, and escalation fears. We then distributed Phases 2 and 3 of each relevant scenario, and both teams jointly discussed the extent to which these new acts of aggression would affect their preferred courses of action, what each hoped the other team would and would not do, and any escalation concerns they had.

Full details on the scenarios are available in Appendix B. The following subsections briefly summarize the scenarios.

Scenario 1

A newly elected Democratic Progressive Party (DPP) government in Taiwan announces plans to accelerate implementation of the ODC and expand nationwide stockpiling of fuel, food, water, medical supplies, and munitions. This includes the delivery of a much higher volume of asymmetric defensive weapons from the United States, requiring the use of civilian merchant ships. As part of a previously announced counterdrug campaign, the CCG stops a Taiwanese freighter traveling from Southeast Asia about 75 miles south of Kaohsiung on Taiwan's side of the Taiwan Strait median line and demands that it submit to a search for contraband. The Taiwanese vessel, which is carrying U.S. arms and other supplies for Taiwan's strategic stockpile, refuses to stop, prompting a warning shot from the CCG ship. The shot goes awry and strikes the Taiwanese ship, sinking it. The CCG vessel rescues and arrests the Taiwanese crew, who are taken back to China for prosecution. Beijing then announces that all ships from Myanmar, Laos, Vietnam, and Thailand must stop in a Chinese port for inspection before proceeding to Taiwan, and the CCG will monitor all maritime traffic in the western portion of Taiwan's ADIZ. Companies that do not comply will be punished with the seizure of their assets in China, exclusion from the Chinese market, and possible impoundment of their vessels. As ships from these nations begin switching flags to avoid the new rules, China further escalates the situation by announcing that all vessels traveling to Taiwan must first stop in a Chinese port for inspection. The CCG impounds several Taiwanese ships.

Scenario 2

A TCG vessel en route to resupply Taiwanese marines on Pratas/Dongsha Island is surrounded by Chinese fishing vessels. A nearby CCG ship radios and demands that the TCG vessel return to Taiwan. Seeking to escape encirclement, the TCG ship rams a Chinese fishing vessel and sinks it. It is then swarmed by more fishing ships, boarded, and seized by the CCG. As Taiwan plans another resupply mission, the People's Liberation Army (PLA) announces exercises in the waters and airspace near where the first TCG ship was taken, including boarding drills and air defense operations against target drones impersonating Taiwanese supply planes. Beijing also announces that it will mine the waters around Pratas. When the Taiwanese government sends a second relief ship, it suddenly loses communications with forces on the island. The PLA Navy states that it has a hospital ship on hand and offers to transport Taiwanese troops off Pratas and back to Taiwan, giving China control of the island.

Scenario 3

Two days after a large U.S. congressional delegation departs Taiwan, Chinese drones overfly Matsu, Penghu, and Wuchiu Islands, releasing a stream of photos of Taiwanese military installations there on social media. Two days later, a single J-20 fighter overflies Kinmen and another overflies Matsu. Within a week, China announces a new East China Sea ADIZ to overlap Taiwan's ADIZ completely, a move made necessary by political air threats that include visits by foreign officials masquerading as civilian flights. The PLA Air Force also begins flying drones around Taiwan's main island, venturing into waters just off the island's southern and eastern coasts. The Chinese news outlet *Global Times* further escalates the conflict by calling for a change to China's No First Use (NFU) nuclear doctrine and printing interviews with Ministry of National Defense officials who suggest that any incursion into China's new ADIZ should be met with the severest of consequences.

Chapter 3. Tabletop Exercise Results and Analysis

This chapter presents general takeaways from across all three TTX scenarios. The following are among the main takeaways:

- Escalation from gray zone to unintended war was highly unlikely, because of U.S. cost avoidance, Taiwanese dependency on U.S. support, and Chinese escalation dominance.[15]
- Blue was significantly more willing and provided more support when Taiwan had already taken steps Blue thought would bolster Green's defense.
- The United States and Taiwan differed in their view of the conflict's stakes. Both viewed existential threat (e.g., blockade, invasion) as their redline for military escalation, but Green rated sovereignty challenges as more threatening than Blue.
- Green was often disappointed by the lack of specificity in Blue's promises of support and international pressure.

The sections in this chapter will (1) explore the perception of limited escalation risks, (2) contrast Blue and Green's differing values of the conflict's stakes and redlines, (3) outline each side's proposed responses (they generally agreed on the need for greater coordination and avoiding escalation, but Green demanded greater clarity and assurance from Blue, which typically demurred), and (4) identify the key factors contributing to escalation risk and the relative magnitude of each factor.

In drawing conclusions from our SME discussions and TTX, we noted several caveats. While we tried to include a variety of experts in our TTX, our exercise inevitably included only a relatively small number of researchers. Because they reached conclusions through discussion, more-articulate or forceful players might have had an outsized influence on the group. A different set of experts might have arrived at different conclusions. We attempted to mitigate these limitations by recording any disagreement in or between the teams. Furthermore, our analysis was firmly grounded in the unique context of the Taiwan dispute. While it might produce some insights for gray zone conflict and escalation in other regions, more research will be needed to determine whether these conclusions apply to any other conflict.

Escalation Risk

Most importantly, escalation of gray zone challenges to conventional warfare was unlikely because of three factors. First, Blue did not perceive the gray zone to be critically important. The team had a clear redline: The United States would only respond militarily to existential threats to

---

[15] To clarify, the scenario assumes that Beijing has not decided to use a gray zone altercation as a pretext for invasion. That was a key concern for the TTX participants; were invasion preparations detected, both Blue and Green would have been likely to escalate.

Taiwan, i.e., actions that would force the Taipei government to acquiesce to Chinese Communist Party (CCP) control over the island. While any scenario that threatened to give the CCP such control was felt to merit military intervention, participants generally believed anything short of a large-scale amphibious invasion or near complete cutoff of vital food and energy imports would fail to compel Taipei to surrender its de facto independence. Short of that, however, Blue did not feel U.S. interests were sufficiently engaged to offset the costs of Chinese retaliation.

In fact, one TTX participant raised the possibility that China's gray zone provocations might serve as a political release valve. These actions may allow Beijing to mollify CCP hardliners even though these actions are generally ineffective at coercing Taiwan and increase Taiwanese hostility and self-identification. If this approach is correct and if the United States and Taiwan become more effective at responding to these provocations, China might then resort to higher escalation.

Second, while it pushed for more-robust U.S. support, Green was unwilling to let its response get significantly ahead of Blue. The team was very clear about Taiwan's relative military inferiority and inability to keep pace with most Chinese escalation. Consequently, the team's plans were contingent on U.S. support. When Blue vetoed more aggressive responses, Green did not take unilateral action.

Third, participants feared that escalation would only benefit Beijing, which enjoys military superiority in the region. Blue stated that even the United States had few cost-effective options in the gray zone, leaving escalation to conventional military force as the only means of rolling back Chinese incursions. In general, Blue felt it possessed no good, symmetric options and instead sought to de-escalate Beijing's challenges to buy time to build Taiwanese and U.S. defenses. Moreover, the United States team did not feel that any of the issues at stake provided decisionmakers with sufficient benefits for escalation. Team members were particularly pessimistic about Taiwan maintaining control of Pratas Island in Scenario 2, considering it indefensible and not covered under the Taiwan Relations Act. Note that this does not represent official U.S. policy on the matter, only the views of the experts on our Blue team.

Contrasting Redlines and Stakes

As stated, Blue's redline for escalation to conventional military force was an existential threat to Taiwan. The team much preferred Green to focus on those threats, accepting the costs of uncontested gray zone incursions as an additional cost of defending against future invasion.

Of the scenarios, Scenario 1 got closest to crossing the U.S.' redline. Chinese economic control or a soft blockade would clearly strangle Taiwan economically and eventually force capitulation. However, Blue still did not respond militarily because the team felt Beijing's actions would galvanize international support for Taiwan, leading to Chinese isolation. In addition, because the scenario specified that Taipei was implementing the ODC, it lengthened the window for international assistance to arrive and financial pressure on Beijing to work.

Importantly, Blue was relatively relaxed even in this situation because Chinese actions were not accompanied by a military build-up necessary for an invasion. The team indicated that, were such preparations detected, they would take the situation much more seriously and be significantly more likely to go beyond intelligence sharing and deploy U.S. military assets to the area.

Across the scenarios, Green did not establish a clear redline because the team's actions hinged on Blue's strategy, support, and risk acceptance. At least within the gray zone, Taipei did not feel like it could respond on its own.

There was a clear disagreement between the United States and Taiwan on the stakes involved. Blue tended to isolate each situation, treating it as a one-off incident that did not augur any greater Chinese plan. The team was not concerned about closing international space or sovereignty concerns held by Green. In fact, it felt Chinese gray zone aggression could increase Taiwan's international space and the willingness of other nations to deal with Taipei. Regardless, while Taiwan's continuing existence constituted a U.S. interest, these other issues did not.

By contrast, Green saw the incidents as interconnected, linked to a longer history and wider set of Chinese actions to foreclose Taiwan's foreign policy and constrain its domestic policy. In the team's view, China would only undertake the scenarios' actions if it had a broader plan supporting them (this view was shared by SMEs in Taiwan and INDOPACOM). Without this supporting plan, the international reaction Green expected would make these steps prohibitively costly for Beijing, even as it would successfully put pressure on Taiwan.

Consequently, Green was particularly concerned that an initial failure to respond forcefully would embolden Beijing to further limit Taiwan's sovereignty. This also drove the team to ask Blue for political signals to address this aspect of China's challenge. In fact, one Blue participant worried that Green was asking for certain responses because of their political effects rather than their military utility.

To be clear, like Blue, the Taiwan team primarily selected responses based on their security impact. But there was some disagreement between Blue and Green on the relative importance of these political objectives. Green was much more concerned about normalization of Chinese actions and control, eroding Taiwan's medium-term ability to resist Beijing's pressure, sustaining international political support, and maintaining regular trade and diplomatic contacts.[16] Green was particularly interested in the U.S. response strategy if Chinese actions ended up not being one-off incidents. However, Blue was often able to demure and avoid specific commitments, contributing to Green's concerns.

---

[16] Experts in Taiwan and INDOPACOM shared this concern about normalization, prompting a need for stronger initial responses to deter future coercion. INDOPACOM experts raised the possibility of Beijing announcing a *principle*—which might or might not be enforced—to lay the groundwork for increasing pressure on Taipei. They cited the soft blockade covered in Scenario 1 and China's regular exercises *de facto* control over the areas around the outer islands. Alternatively, Beijing could disrupt resupply to Pratas (as in Scenario 2) or in some way regularize its jurisdiction over the island.

Proposed Responses

Across the scenarios, Taiwan pushed for greater clarity of U.S. redlines, U.S. leadership in coordinating an international response, and enhanced military cooperation and technology-sharing. The United States largely rejected these, hoping to de-escalate tensions and restrain the Taiwanese response. The team was concerned about intentionally or inadvertently giving Taiwan a blank check. In Scenario 1, the team repeatedly sought clarification on Green's Rules of Engagement and encouraged a more muted response in public statements.

Notably, Green never proposed attacking Chinese military, coast guard, or even maritime militia forces. The team's posture was fundamentally reactive and deliberately avoided a violent response, although it did leave open the possibility of returning fire if clearly attacked.

Blue was also concerned about the political and morale effects if a Taiwanese military response or demonstration failed. For example, in Scenario 3, the United States worried that a missed shootdown of a Chinese fighter would embarrass Taipei, crater Taiwanese morale, embolden Beijing, and escalate the situation for no gain.

Blue attempted to align Taiwan's interests and policies with U.S. strategy. Most importantly, when the scenario specified that Taiwan had not implemented the ODC, the United States encouraged it to do so. Notably, Scenario 1 states that Taipei had been implementing the strategy prior to the situation's start. Several Blue team members highlighted that this made them much more inclined to provide support. In essence, the United States rewarded greater Taiwanese effort in their own defense. Similarly, the United States encouraged Taiwanese cost acceptance, focusing on the existential threat, and ignoring challenges below this level. This included giving up Pratas Island if attacked and accepting increasing encroachment of Taiwanese airspace and maritime zones.

Across the scenarios, the teams settled on three recurring courses of action. First, both wanted to avoid militarizing the situation. However, Green considered deploying TCG vessels in Scenarios 1 and 2, which Blue thought might tempt China into further demonstrations of Taiwan's inability to respond to escalating challenges.

Second, Green pressed for and Blue shared intelligence to gain a stronger picture of Chinese operations and deployments. One participant observed that the scenarios revealed the need for reliable, redundant communication channels between Taiwan and the United States at the national (not tactical) level to facilitate understanding during crises. The timing of when to introduce these channels is also important; participants stated that they should be tied to Chinese actions to reinforce the consequences of Beijing's misbehavior. More broadly, the TTX participants puzzled over how to get China to internalize the link between Beijing's provocations on the one hand and Taiwanese defense preparations (especially the ODC) and U.S. support on the other hand.

Third, both teams hoped to internationalize each scenario, highlighting the transgressive nature of Chinese actions to garner greater foreign support for Taiwan's position. The United

States viewed these situations as opportunities for Taipei and Washington to paint Beijing as a bad actor and maintain the overall trend of growing international support for Taiwan.

However, Green was often disappointed with the vagueness of Blue's assurances about international support. The team pushed the United States to lead collective, non-Taiwanese efforts to condemn Chinese actions and coordinate responses. Green frequently pressed the United States for details on how international positive feelings for Taiwan would translate into concrete material support. This was particularly true of Scenario 1, where state and commercial attempts to run the soft blockade were essential to Taiwan's economic livelihood. Blue was willing to provide public statements of support and encourage other countries to do so as well. This, it felt, would better position Taiwan, the United States, and regional partners in the event China tried the same gray zone tactic again. But the United States stopped short of the more-concrete coordinating activities Green preferred. When asked directly whether Blue thought other countries would act if the United States did not, the team was more concerned about the escalation risk and overpressurizing the situation, as well as buying time to better respond to existential threats.

Factors Affecting Escalation Risk

The TTX participants identified several factors affecting the risk of escalation and their relative magnitude. Table 3.1 summarizes these, dividing them into their effects on Taiwanese versus U.S. escalation potential. Because of U.S. constraints on Taiwan, the overall escalation potential matches the U.S. column.

Table 3.1. Escalation Risk and Magnitude

Gray Zone Challenge	Escalation with Taiwan	Escalation with the United States	Overall Escalation Potential
High expected degree of Chinese control	High	High	High
Close to Taiwan's main island	High	High	High
Use of military forces	High	Moderate	Moderate
Nuclear threats	Low	Moderate	Moderate
Sovereignty challenges	Moderate	Low	Low
Use of non-conventional forces	Moderate	Low	Low

Both Taiwan and the United States considered threats with a high expectation of Chinese control of Taiwan (e.g., invasion, economic control) and those close to the main island (including the Penghus) to warrant military escalation. But they differed on the importance of sovereignty challenges and nuclear threats. The former was previously discussed. However, Taiwanese SMEs did not think China would use nuclear weapons against Taiwan. Beijing wants to rule the island, not destroy it. In the SME's view, nuclear threats aim to prevent outside intervention, specifically from the United States and Japan. That drove a higher U.S. estimate of escalation risk, even as project participants considered the likelihood of nuclear saber-rattling to be low.

Chapter 4. Conclusions, Implications, and Recommendations

This chapter compiles additional policy implications and future research that emerged from the project. The first section explores changes to the TTX scenarios that will lead to more-escalatory and challenging situations for Blue and Green. The second section highlights the additional scenarios proposed by the SMEs and TTX participants. The third section explores the two foundational issues on which the United States and Taiwan differ. The TTX participants and SMEs agreed on many aspects of China's gray zone challenge and the limits under which Taiwan and the United States operate. They differed, however, over Taiwanese (1) hedging and (2) moral hazard.

Future Research

Several areas for future research emerged from the TTX. First, neither Blue nor Green possessed nonescalatory options that would have rolled back Chinese incursions. Taiwanese concerns about how international support would translate into concrete action and the lack of U.S. clarity on this point reinforced the limited availability of responses. Follow-on research could explore the capabilities and technology Taiwan and the United States could develop to symmetrically meet Chinese actions and avoid escalation. This would be particularly useful for Scenario 1, which the participants found the most threatening and difficult.

Second, the participants found each individual scenario to be manageable, although Green had to deal with both China's actions and U.S. constraints. The world *reset* across scenarios, meaning that Blue, Green, and Red actions in one scenario did not affect subsequent situations. However, several participants stated that, were the scenarios combined or not reset, they would find China's actions much more escalatory and challenging. While the scenarios never prompted a U.S. military response individually, they might if run cumulatively. Future research could explore how escalation dynamics and strategic responses change if the scenarios are combined.

Third, both Blue and Green's strategies focused on raising international support for Taiwan. But would other countries respond to the scenarios in the way the TTX participants expected? Future research could run the scenarios—individually or cumulatively—but include participants representing such countries as Japan, Australia, and the Philippines to gauge the effect of Chinese actions on third-party support for Taiwan. Of particular interest, the U.S. team assumed that Taiwanese restraint would bolster foreign views of Taipei. But is that the case? Similarly, how would these other countries view U.S. action or lack of action in prompting their own responses and for preexisting U.S. commitments to their security?

Fourth, Blue worried that a failed Taiwanese military response could lead to a variety of political blowback for Taipei, including reduced domestic morale and decreased international support. However, Gelpi, Feaver, and Reifler suggested that democratic publics are willing to accept casualties and military inefficacy so long as a viable path to victory exists.[17] A failed military mission might bolster Taiwanese public support for resistance if it can nevertheless demonstrate a credible warfighting plan. But future research can answer this question more precisely: What kinds of mistakes and successes affect Taiwanese will to fight? And what has a greater effect on Taiwanese morale: ignoring gray zone coercion or a failed response?

Fifth, beyond Taiwan, China uses gray zone coercion against several other Asian countries. Do these states consider the factors contributing to escalation for Taiwan (e.g., proximity, expected degree of Chinese control, etc.) to be equally dangerous for them?

Additional Scenarios

TTX participants and Taiwanese and INDOPACOM experts suggested several other gray zone challenges that could be featured in future research. We had previously considered some—such as cyberattacks on Taiwanese critical infrastructure and mis-, dis-, and malinformation campaigns—but cut them in favor of our three TTX scenarios, which RAND and Taiwanese experts considered both more likely and more escalatory.

In several others, the U.S. Department of Defense would have limited or no supporting role in either U.S. or Taiwanese responses. These scenarios included hostage-taking, election interference, and political assassination. But experts also considered these to be among the most challenging scenarios, in part because China and Taiwan have established rules to manage military-to-military and coast guard–to–coast guard interactions.[18] Instead, Beijing could work with local Taiwanese politicians or gangs to present law enforcement challenges to Taiwan's central government that simultaneously have significant national security implications.

For example, one expert noted that the CCP has deep connections with local politicians on Jinmen. Beijing might be able to pressure these individuals into leading public protests against the local military garrison. They would paint the installation as provoking China, demand its removal, and escalate to cutting off water supplies and electricity. Such an operation would attack the seams in Taiwan's domestic security structure, which does not have an overarching authority similar to the U.S. Department of Homeland Security. Instead, Taiwan's Defense Ministry would be legally unable to respond, and the police would follow the local politicians' authority. Taiwan would, in effect, be blockading itself. If the garrison were removed, however, this would reduce Taiwan's military readiness and intelligence collection on China.

Similarly, China could encourage and support Taiwan-based mafia or gangs to seize control of and occupy energy infrastructure (e.g., a nuclear power plant or electrical station) or a

---

[17] Christopher Gelpi, Peter D. Feaver, and Jason Reifler, *Paying the Human Costs of War: American Public Opinion and Casualties in Military Conflicts*, Princeton University Press, 2009.

[18] To clarify, China and the Kuomintang (KMT) concluded these agreements. Beijing has not necessarily adhered to them with the Tsai administration.

transportation node (e.g., a harbor or high-speed rail station). These groups might outgun Taiwanese police, but, despite the homeland and national security implications, the military may not be able to legally respond.

U.S. experts at INDOPACOM and in the TTX also highlighted the following:

- Chinese dredging off Jinmen to build an airport, restricting what can land and take off from there (likely, but not as dangerous as other scenarios)
- regulatory pressure on Taiwanese businesses or third-party nationals who have been to Taiwan (likely, but not as dangerous as other scenarios)
- a natural disaster causing significant damage to Taiwan, with Beijing (1) taking advantage of the temporarily reduced communications with Taipei to route relief assistance through Chinese ports, (2) further regulating outside contact, or (3) monopolizing such efforts itself and denying access to other countries (dangerous, but depends on opportunity)
- rolling *virtual blockades* using military exercises and missile firings to selectively close maritime zones
- using the Maritime Militia and CCG to close fishing areas important to Taiwan for ecological or other reasons.

SME-TTX Differences

The SMEs and TTX participants mostly agreed on the dynamics of Chinese gray zone challenges and both Taiwanese and U.S. constraints. However, the groups perceived two foundational issues differently. First, the TTX participants expressed frustration that Taipei was not sufficiently implementing the ODC or any other asymmetric defense concept. The SMEs, by contrast, pointed to the lack of U.S. commitment as the major impediment. Overall, Taiwan and the United States are trapped in a prisoner's dilemma: Both would benefit from aligning their security strategies, but each has incentives to wait for signals from the other side before committing.

Second, the TTX participants (the Blue team in particular) worried about moral hazard issues: U.S. support might embolden Taipei into rash action. But the SMEs expected Taipei to respond cautiously to any provocations, conditioning its response on U.S. direction. This section explores these perceptual differences in greater depth. The analysis suggests that Washington has greater constraining power over Taipei than the TTX participants believed and, therefore, might be overestimating the risks of Taiwanese entrapment and adventurism.

In drawing conclusions from our SME discussions and TTX, we would like to reiterate several caveats. While we tried to include a variety of experts with experience in various fields of foreign policy, the TTX inevitably included only a relatively small number of researchers. Because they reached conclusions through discussion, more-articulate or forceful players might have had an outsized influence on the group. A different set of experts might have arrived at different conclusions. We attempted to mitigate these limitations by recording any disagreement within or between the teams. Furthermore, our analysis is firmly grounded in the unique context of the Taiwan dispute. While it might produce some insights for gray zone conflict and escalation in other regions, we caution that more research will be needed to determine whether these conclusions apply to any other conflict.

Hedging and the Prisoner's Dilemma

There was disagreement among the Taiwanese SMEs about how to manage relations with China, both specifically regarding Beijing's gray zone challenges and more broadly. This tended to break along party lines; DPP-affiliated experts were more likely to advocate greater resistance to Chinese activities, while KMT-affiliated experts were more likely to seek accommodation and proactive management or coordination of the Taiwan-China relationship. The latter pointed to a series of agreements signed between the prior KMT Ma Ying-jeou administration and Beijing as evidence of the efficacy and advisability of this approach.

This division highlights two points in U.S. policy discussions about Taiwan. First, Taipei has significantly more foreign policy options than U.S. experts assumed. For example, in plenary discussion, the TTX's Blue team raised the possibility of the United States abandoning Taiwan if the latter did not make sufficient defense efforts, specifically and fully implementing the ODC. But there was no discussion about how greater Taiwanese accommodation of China on gray zone challenges and broader security and economic issues would affect U.S. interests. Even when raised, this Taiwanese option was not deeply considered as a possibility.

Second, Taiwanese uncertainty about U.S. commitment drives this policy division. Many experts felt that the lack of commitment buttresses the accommodationist approach. Politicians or commentators could more justifiably claim that Taiwan was on its own, necessitating dialogue and compromise with Beijing.

Most importantly, this U.S. problem in Taiwanese politics affects the island's defense strategy and force acquisition. Multiple SMEs affirmed that successful defense requires U.S. intervention and requires Taipei to purchase high-end and expensive weapons systems, in part because they are a concrete signal of U.S. support. The ODC, by contrast, lacked this political symbolism, reducing the perceived likelihood of U.S. intervention. Moreover, that asymmetric strategy would buy time but not victory in and of itself. The SMEs expected U.S. intervention would still be required even if the ODC were implemented fully.

Taiwan and the United States are trapped in a prisoner's dilemma. Both sides would benefit—individually and collectively—from improved alignment of their security strategies and defense acquisition. But their strategies hinge on one another's: Both sides are waiting for a signal of commitment from the other before deeper cooperation. On one side, the United States wants Taiwan to fully implement the ODC to demonstrate its seriousness of purpose. As evidenced by the Blue team's response, Washington might provide greater security and political support.

On the other side, Taiwan is waiting on a clear U.S. defense commitment before accepting the domestic political risk of implementing the ODC, particularly against Ministry of National

Defense slow-rolling. The country also worries about U.S. abandonment (a possibility Blue brought up), which would negate Taiwanese defense preparations while raising tensions with China.

Consistent communication can overcome the prisoner's dilemma, although this would require the U.S. executive and legislative branches to coordinate their foreign policies. Analysts highlight the need for conditionality, tying increased U.S. support to Taiwan meeting specific benchmarks.[19] However, those benchmarks must be clearly specified and coordinated, and the United States must reassure Taipei about the security guarantees it can expect for fulfilling those conditions. Some Taiwanese SMEs noted past United States moving of goalposts. Others disputed the notion that Taiwan was not doing enough for its own defense, emphasizing that Taiwan spends a greater percentage of its national budget on defense than the United States does.

Moral Hazard

The second difference between TTX participants and SMEs concerned moral hazard. As mentioned, the TTX Blue team worried about inadvertently giving Green a blank check in responding to Chinese provocations. Providing excessive support could embolden Taiwan into provoking Beijing, believing it had greater U.S. backing than it possessed. Blue focused much of its efforts on shaping Taiwan's response, rather than addressing China's actions independently or in coordination with Green. Blue was also keen to avoid a situation in which Taiwan forced the United States to bail it out. One participant raised the fact that Washington could abandon Taiwan without discouraging its allies if those countries felt Taipei had needlessly provoked China. Nevertheless, there was an implicit belief that Washington could not prevent Taipei from adventurism.

This contrasted sharply with the discussions among Green and with the Taiwanese SMEs. Neither group once mentioned independence, and there was a clear perception of Taiwan's dependency on U.S. assistance and military inferiority vis-à-vis China. In fact, Taiwanese SMEs were perhaps more cautious in their proposed responses to the scenarios than INDOPACOM SMEs. Here, there was explicit agreement that multiple, overlapping pressures were actively and effectively constraining Taipei.

Conclusion

Overall, this project found that the risk of escalation from the gray zone to conventional conflict is quite limited. Escalation was believed to benefit Beijing and play to its strengths; the United States was unwilling and unable to effectively contest the gray zone; and Taiwan faced

---

[19] Michael A. Hunzeker and Alexander Lanoszka, "Real Friends Twist Arms: Taiwan and the Case for Conditionality," War on the Rocks, July 27, 2022.

superior Chinese capabilities and U.S. constraints. During the TTX, Blue and Green had significant differences in their perceptions of stakes, preferred responses, and concerns about Chinese escalation. That said, both TTX Green team and the Taiwanese SMEs recognized the necessity of U.S. support, sticking closely to U.S. positions while seeking to bolster the relationship.

Moreover, the project identified several additional areas for future research. First, the TTX scenarios could be combined to examine their cumulative impact on escalation. Second is exploring the network effects of the base scenarios by including additional Green teams. Do Japan, Australia, or additional countries respond in the ways anticipated by this project's Blue and Green teams? Last of all, the SMEs and TTX participants provided additional gray zone scenarios affecting the missions and strategies of the U.S. Department of Defense and other U.S. government entities (particularly the U.S. Department of Homeland Security).

Finally, the project identified two key differences in threat and escalation perceptions between the TTX participants on the one hand and the Taiwanese and INDOPACOM SMEs on the other hand. These drove prisoner's dilemma and moral hazard dynamics between the United States and Taiwan, preventing deeper security cooperation. But the analysis suggested that greater communication and clear policy standards could overcome the prisoner's dilemma, while U.S. experts might overestimate the risk of Taiwanese adventurism.

Appendix A. List of Initial Scenarios

The project team presented the following scenarios to RAND experts, who then rated each on likelihood of occurrence and escalation potential. The first section details potential Chinese gray zone activities, providing a brief description and instances where Beijing has conducted similar operations in the past.

The project team also considered U.S.- or Taiwan- instigated activities that would prompt Chinese escalation. These are presented in the following section. Importantly, RAND experts considered these scenarios to be less likely to occur than the Chinese gray zone ones.

Possible Chinese Gray Zone Activities

Evacuation of citizens: Because of political unrest and civil disturbance in Taiwan, China declares the evacuation of its citizens from the main island. China sends military aircraft to major Taiwanese airports, supported by Chinese naval ships within the strait. In recent years, the PLA has made much of its growing ability to evacuate Chinese civilians and protect them from harm.[20] A crisis endangering PRC nationals on Taiwan could also prove an opportunity for Beijing either to achieve some degree of military access (even if it is unarmed logistical units) to the island or to place the Taipei government in the position of refusing aid to endangered civilians.

UAV or missile overflight: China launches missiles over or flies UAVs directly through Taiwanese airspace (not just the ADIZ). China has greatly increased the size, frequency, and boldness of its military incursions into Taiwan's ADIZ and fired missiles over Taiwanese territory; additional overflights are a distinct possibility, should China feel the need to express its displeasure in the future.[21]

Island seizure: China launches a joint firepower and joint island-landing campaign against Taiwanese-held Pratas Island. China has, in the past, menaced, occupied, and even assaulted islands in the South China Sea.[22] It has also menaced the Taiwanese offshore islands of Quemoy

---

[20] Nathan Beauchamp-Mustafaga, "PLA Navy Used for First Time in Naval Evacuation from Yemen Conflict," Jamestown Foundation, *China Brief*, Vol. 15, No. 7, April 3, 2015.

[21] "China Sends Missiles Flying Over Taiwan," *The Economist*, August 4, 2022.

[22] See Center for Strategic International Studies, undated. Note that, while the PLA Navy has not launched full naval assaults on occupied islands in its more recent history in the South China Sea, it fought battles with South Vietnamese and later unified Vietnamese forces there in 1974 and 1988, resulting in the sinking of ships and dozens of deaths. See Nga Pham, "Shift as Vietnam Marks South China Sea Battle," BBC, January 15, 2014; Drake Long, "Chinese Maritime Militia on the Move in Disputed Spratly Islands," Radio Free Asia, March 24, 2020.

and Matsu and conducted numerous military flights between Taiwan and the Pratas Islands.[23] Seizure of Pratas Island would be a significant escalation of Chinese practices but might not be impossible as tensions continue to mount.

Interference with Taiwan military operations: China captures or sinks a Taiwan supply convoy on the way to Itu Aba. While China generally avoids direct clashes between naval ships, it has repeatedly interfered with the civilian and government vessels of rival claimants in the South China Sea, and these incidents have led to sinkings.[24] Beijing has also nonviolently challenged U.S. naval and air forces operating in the region.[25]

Blockade or quarantine: China announces that, as the only legitimate authority able to legally regulate Taiwanese trade, any commercial ship going into a Taiwanese port without first stopping in a PRC port and operating according to PRC law will be impounded or sunk. Blackwill and Zelikow detailed this scenario in their analysis of critical threats to Taiwan.[26] While Beijing has not sought this level of physical control over Taiwanese trade, the PLA has also extensively discussed and practiced blockades of the island. Beijing has also locked Philippine vessels out of the Scarborough Shoal and prevented Vietnamese ships from interfering with a Chinese oil exploration near its shores.[27]

Harassment or sinking of a Taiwanese civilian ship: Chinese Maritime Militia and CCG vessels harass, ram, and sink a Taiwanese civilian ship at sea. As noted earlier, China has a history of harassing civilian and government ships of rival claimants in the South China Sea.

Cyberattack on infrastructure: Chinese hackers take down the Taiwanese power grid, causing permanent physical damage at some plants and transmission stations. Taiwan faces millions of cyberattacks and intrusions a day, many of which originate from the PRC.[28] China has not yet launched large-scale, physically destructive attacks on Taiwanese infrastructure, but

---

[23] Gerry Doyle, Anand Katakam, Ben Blanchard, and Marco Hernandez, "The Skies over the South China Sea," Reuters, October 20, 2021; Jane Perlez and Amy Chang Chien, "Chinese Drones: The Latest Irritant Buzzing Taiwan's Defenses," *New York Times,* September 10, 2022.

[24] U.S. Department of Defense, "China Coast Guard Sinking of a Vietnam Fishing Vessel," press release, April 9, 2020.

[25] Luis Martinez, "A Look at the U.S. Military's Close Calls with China, Russia in the Air and at Sea," ABC News, October 2, 2018.

[26] Robert D. Blackwill and Philip Zelikow, "The United States, China, and Taiwan: A Strategy to Prevent War," Council on Foreign Relations, Special Report No. 90, February 2021.

[27] Michael Green, John Schaus, Jake Douglas, Zack Cooper, and Kathleen H. Hicks, *Countering Coercion in Maritime Asia: The Theory and Practice of Gray Zone Deterrence*, Center for Strategic and International Studies, May 9, 2017.

[28] Shinji et al., 2023; "Taiwan Government Faces 5 Million Cyber Attacks Daily: Official," AFP, October 11, 2021; Shelley Shan, "Record Number of Cyber Attacks Reported," *Taipei Times*, August 5, 2022.

has already begun to lay the groundwork necessary to launch such attacks in the United States and might have already done so in Taiwan as well.[29]

Political assassination: China assassinates key Taiwanese political figures (e.g., president, vice president, defense minister). This would be highly unlikely and a major escalation from current Chinese practice. That said, given China's oft-stated belief that Taipei's intransigence is the work of a small group of hardcore separatists and its personal vilification of Tsai Ying-wen (Taiwanese president from 2016 to 2024), we felt that this scenario merited consideration.[30] Following the murder of former Japanese prime minister Shinzo Abe, a nationalist television pundit appeared to justify the potential assassination of the Taiwanese president.[31]

Economic coercion: China nationalizes Taiwanese companies' assets on the mainland, halts critical Chinese exports to Taiwan, and embargoes Taiwanese exports to China. Beijing has used its "sharp power" to retaliate against foreign companies and countries for failing to comply with its policies or framing of political issues.[32] Sharp-power tactics include an unofficial embargo against South Korea for allowing the United States to deploy the Theater High Altitude Air Defense System system there and ban on Australian beef and wine.[33]

Hostage taking: China detains Taiwanese citizens without coordinating with Taipei. This could either be through interfering with extradition processes (China demands that extradited Taiwan citizens are first sent to China), preventing Taiwan citizens in China from leaving, or kidnapping Taiwan citizens in Taiwan. In recent years, Beijing has insisted on the extradition of hundreds of Taiwanese citizens to the PRC.[34] The country also has a history of kidnapping and detaining those it considers threatening.[35] Most famously and recently, China detained two

[29] Nicole Perlroth and David E. Sanger, "China Breached Dozens of Pipeline Companies in Past Decade, U.S Says," *New York Times*, July 26, 2021.

[30] "Diehard 'Taiwan independence' Separatists to Be Punished According to Law: Spokesperson," Xinhua, November 5, 2021; "The Taiwan Problem and the Project of Chinese Reunification in the New Era" ["台湾问题与新时代中国统一事业"], Qiushi [求实], August 10, 2022; Gao Yang [高杨], "The Will of the People Is Irresistible" ["民意不可违 大势不可挡"], Quish [求实], June 22, 2019.

[31] Keoni Everington, "China Commentator Says Taiwan's President Should be Assassinated," *Taiwan News*, July 15, 2022; Nicolle Liu and Phelim Kine, "China's Internet Expresses Glee at Abe's Assassination," *Politico*, July 8, 2022.

[32] Christopher Walker, "What Is 'Sharp Power?'" *Journal of Democracy*, Vol. 29, No. 3, July 2018; Christopher Walker, "Rising to the Sharp Power Challenge," *Journal of Democracy*, Vol. 33, No. 4, October 2022.

[33] Florence Wen-Ting Yang, "Asymmetrical Interdependence and Sanction: China's Economic Retaliation over South Korea's THAAD Deployment," *Issues and Studies*, Vol. 55, No. 4, 2019; Kim Jiyoon, John J. Lee, and Kang Chungku, "Changing Tides: THAAD and Shifting Korean Public Opinion Toward the United States and China," ASAN Institute for Policy Studies, March 20, 2017.

[34] Safeguard Defenders, "China's Hunt for Taiwanese Overseas," fact sheet, 2021.

[35] Tom Phillips and Oliver Holmes, "Activist Who Vanished in Thailand is Being Held in China, Says Wife," *The Guardian*, February 3, 2016.

Canadian individuals for nearly three years in retaliation for Canadian compliance with a U.S. extradition request for a Huawei executive.[36]

Nuclear tests: China detonates a nuclear device in the Pacific Ocean east of Taiwan. While we considered this scenario improbable, Beijing stated that it possessed a neutron bomb in 1999 and obliquely hinted that it was willing to use it against Taiwan.[37] Further, Beijing has increased its nuclear arsenal over the past year, which could indicate a change in doctrine.[38] Finally, we believed this scenario was worth consideration despite its improbability because it could shed some light on the ways in which weapons of mass destruction could affect escalation dynamics around Taiwan.

Interdiction of weapon transfers: China interdicts shipments of advanced U.S. weapons to Taiwan. Beijing has previously impounded Singaporean ground vehicles and publicly objects to every U.S. arms sale to Taiwan.[39]

Forced refugee or prisoner transfers: China deports prisoners to Taiwan on the pretext that they are transferring detainees within Chinese territory. Again, highly unlikely, but China has transferred populations in Xinjiang and Tibet.[40]

Possible Taiwanese or U.S. Activities

Stationing of U.S. troops on Taiwan: Taiwan invites the United States to station one to three combat brigades and needed enablers on the island, and the United States accepts. The United States once had a large military presence on the island, before normalization of relations with China. Since then, it has maintained a very small but growing contingent of soldiers there.[41] Rapid growth of this force is unlikely but may not be impossible should the United States or Taiwan feel it necessary to put U.S. personnel into harm's way to deter China.

Restart nuclear weapon program: the Taiwanese government rebuilds its reprocessing facilities, prompting an International Atomic Energy Agency announcement that they have started a nuclear weapon program. The Taiwanese government has sought to develop nuclear weapons in the past.[42] Should Taipei doubt Washington's ability or willingness to defend the island, a nuclear weapon program might be Taiwan's only hope of maintaining its independence.

---

[36] "Meng Wanzhou and the Two Michaels: A Timeline," *The Guardian*, September 24, 2021.

[37] Seth Faison, "China Proclaims It Designed Its Own Neutron Bomb," *New York Times*, July 15, 1999.

[38] U.S. Department of Defense, *China Military Power Report*, November 29, 2022.

[39] "China Paper Says Singapore Troop Carriers Should Be 'Melted Down,'" Reuters, November 28, 2016.

[40] Austin Ramzy and Chris Buckley, "'Absolutely No Mercy': Leaked Files Expose How China Organized Mass Detention of Muslims," *New York Times*, November 16, 2019; Tsultim Zangmo, "The New Trends of Han Migration to TAR," *Tibet Journal*, Vol. 44, No. 1, 2019; Emily Yeh, *Taming Tibet*, Cornell University Press, 2013.

[41] Gordon Lubold, "U.S. Troops Have Been Deployed in Taiwan for at Least a Year," *Wall Street Journal*, October 7, 2021.

[42] Nuclear Threat Initiative, "Taiwan Overview," fact sheet, May 27, 2015.

Anti-China educational curricula: The Taiwanese government initiates a new mandatory educational program stridently asserting an independent Taiwanese identity and refuting any cultural or national attachment with the mainland, which is described by some proponents as explicitly meant to create a Taiwanese generation with no ties to China. The CCP already believes that the Taiwanese DPP (in power as of 2023) is set on separating Taiwan from China, in part by actively promoting a non-Chinese identity among the island's people.[43] Since the 1990s, Taiwanese educational materials have given greater emphasis to Taiwanese identity.[44]

Taiwanese cyber hack of Chinese infrastructure: Taiwanese government–affiliated hackers launch a cyberattack that shuts off power in Beijing and causes serious damage to a powerplant there. While Taiwan has not thus far demonstrated a willingness to use such tools, it has itself been a frequent target of cyberattacks and could seek to retaliate in kind.

Summary

Some of these scenarios represent significant escalation from current Chinese practice. While China is certainly interested in redefining the status quo around Taiwan (or at least responding to prevent its redefinition by the United States and Taipei, as Beijing sees it), it generally does so cautiously.[45] In confrontations with its regional neighbors, it has preferred incremental aggression and careful control to prevent spiraling escalation.[46] When it does escalate, China usually does so in ways and areas in which it can avoid direct military confrontation. Even when the actions of individuals have resulted in unanticipated crises, Beijing has generally sought less escalatory measures to achieve its aims. For example, after a Chinese fisherman, possibly acting without government authorization, rammed a Japanese vessel and was arrested, China opted for easily reversible economic actions to secure his release (and prevent Japan's assertion of the right to enforce the law in disputed water) rather than more direct, physical confrontation. Likewise, after a Chinese pilot accidentally collided with a U.S. intelligence aircraft, China demanded an apology from Washington but again avoided major military escalation. Finally,

---

[43] Gao Yang [高杨], 2019.

[44] Meihui Liu and Li-Ching Hung, "Identity Issues in Taiwan's History Curriculum," *International Journal of Educational Research*, Vol. 37, Nos. 6–7, 2002.

[45] Note, for example, that despite Beijing's desire to respond forcefully to Pelosi's visit, it chose to engage in military maneuvers which, while concerning and dramatic, were more political than operational, and China has generally refrained from more-direct interference with U.S. or Taiwanese military units. Furthermore, the CCP has been careful to ensure that sanctions are targeted to cause pain for some of Tsai Yingwen's supporters and has chosen not to take more dramatic and costly action that could devastate Taiwan at the cost of seriously disrupting the PRC. See Bonny Lin, Taylor Fravel, Cristina Garafola, Kathrin Hille, Roderick Hill, and Chris Twomey, "The Military Dimensions of the Fourth Taiwan Strait Crisis," transcript of virtual panel, Center for Strategic and International Studies, August 23, 2022; Brad Lendon, "U.S. Sends Two Warships Through Taiwan Strait, in First Transit Since Pelosi Trip," CNN, August 28, 2022; Mike Ives, and Zixu Wang, "Mostly Bluster: Why China Went Easy on Taiwan's Economy," *New York Times*, August 12, 2022.

[46] Lin, Fravel, et al., 2022.

during a standoff with Vietnam in the South China Sea, China withdrew its militia and other vessels.[47] So long as China continues to follow such practices, uncontrolled escalation to armed conflict is significantly less likely.

That said, China has grown far more aggressive in the past decade and a half, and it is possible that it will continue to do so. Beijing views developments in Taiwan with growing unease, especially the successes of the DPP and strengthening ties between Taiwan and other nations. As China's confidence and its sense of crisis over the threat of a permanent rift with Taiwan grow, significant escalation cannot be ruled out. We believe that these scenarios are sufficiently likely to merit further consideration.

---

[47] "Japan Frees Chinese Boat Captain amid Diplomatic Row," BBC, September 10, 2010; Naval History and Heritage Command, "EP-3 Collision, Crew Detainment, Release, and Homecoming," webpage, August 18, 2021.

Appendix B. Scenarios for Tabletop Exercise

Scenario 1

Part 1

The DPP wins the 2024 Taiwanese elections. The new president, backed by a firm majority in the Legislative Yuan, accelerates implementation of the ODC. Taiwan expands the nationwide stockpiling of fuel, food and water, medical supplies, and munitions, with a goal of five months of self-sufficiency in the event of a full blockade. Under the Taiwan Policy Act, Taipei coordinates with the U.S. Foreign Military Sales program to prioritize the mass delivery of porcupine defense weapons.[48] The DPP unveils a public tracker measuring its progress in meeting stockpiling goals. The government is particularly focused on acquiring these supplies from Southeast Asia, bolstering the New Southbound Policy.

Shortly before this initiative, Beijing launched an antidrug campaign, tasking the CCG with interdicting maritime drug shipments. In this scenario, the CCG identify a Taiwanese civilian freighter possibly carrying contraband. The freighter is approximately 75 miles southwest of Kaohsiung and within the Taiwanese side of the Taiwan Strait median line. At that location, a CCG ship demands that the freighter stop to be searched for drugs onboard.

The Taiwanese vessel refuses and continues sailing, because it is carrying a substantial quantity of U.S. weapons for the stockpiling effort. After two additional attempts, the CCG ship fires a warning shot from its deck gun. The shot is poorly aimed, however, clipping the freighter, which begins taking on water. The crew abandons ship, and the CCG vessel rescues them and takes them to China.

Beijing blames the Taiwanese crew members for the situation and announces that it will prosecute them for violating Chinese maritime law. It announces that the CCG will more aggressively inspect vessels transiting the Taiwan Strait to interdict all contraband (not just narcotics). Beijing also demands that the Taiwanese government provide transit details on all shipments of military goods to the island, to prevent misunderstanding.

Part 2

Beijing steps up its efforts, announcing a new set of maritime rules. It declares that ships coming from Taiwan are subject to heightened inspection in Chinese ports, slowing manufacturing relying on cross-strait trade. Furthermore, commercial ships going from Myanmar, Laos, Vietnam, and Thailand to Taiwan are now expected to first stop in a Chinese

---

[48] "What Is Taiwan's Porcupine Defence Strategy?" *The Economist*, May 10, 2022.

port for inspection. The CCG will monitor traffic in the maritime area roughly overlapping with the western portion of Taiwan's ADIZ.

The rules outline possible penalties for noncompliance. These include various fines, criminal prosecution of crew, temporarily or permanently barring violators from shipping goods to or from the Chinese market, freezing or seizing assets in China, and impounding vessels. Beijing justifies its actions by claiming that, according to United Nations General Assembly Resolution 2758, it is the internationally recognized government of Taiwan. It is simply exercising its sovereign right to regulate the province's trade, counteract weapon proliferation, and assist the global fight against drugs.

Part 3

Shipping companies begin switching flags to avoid having to submit to Chinese inspections. Frustrated and hoping to normalize its regulation of Taiwanese trade, Beijing announces that all vessels going to Taiwan must first stop at a Chinese port. The CCG impounds several Taiwanese ships, seizing medical supplies and oil and lubricants for small arms. Their crews are being tried for violating the new Chinese maritime rules.

Scenario 2

Part 1

In late 2023, a TCG vessel is en route to resupply Taiwanese marines on Pratas/Dongsha Island. At about 200 miles away from Kaohsiung, the ship has passed several fishing vessels and detects about 25 more ahead of it, spread throughout its sailing path. It has also detected a CCG cutter shadowing its movements from several miles away.

The fishing vessels suddenly begin converging on the TCG ship's position, blocking a direct path to Pratas Island. The CCG ship also moves to within one mile. As the fishing boats start surrounding the TCG ship, the CCG vessel radios it and demands that it return to Taiwan. The Taiwanese ship attempts to avoid encirclement and is forced to ram an encroaching fishing boat. That vessel sinks.

The other boats eventually manage to bracket the TCG ship. The CCG ship's crew boards and impounds the vessel. All the fishermen in the sunken boat are rescued. The Taiwanese ship and crew are being transported to China.

Part 2

As Taiwan prepares another resupply mission, the PLA Navy announces a snap exercise in the area where the TCG ship was seized. Using destroyers and helicopters, the PLA Navy simulates the interdiction of several transport ships and their seizure by heliborne marines. Separately, it conducts live-fire practices using antiair systems against dummy drones. The drones are broadcasting signatures similar to those of C-130 aircraft.

In addition, the PLA Navy announces that it is mining the waters around Pratas Island. The areas mined fall within 12 nautical miles of the island, considered Chinese territory. Such actions within that territorial sea comply with the relevant international laws for peacetime mining.

Part 3

As Taiwan's new resupply mission makes way, Taiwan's central military command can no longer communicate with its forces on Pratas.

The PLA Navy privately communicates to the Taiwanese government that the navy have a hospital ship near Pratas ready to provide humanitarian assistance for the Taiwanese marines and others on the island and transport back to Taiwan.

Scenario 3

Part 1

In 2026, a large U.S. congressional delegation visits and departs Taiwan. Two days later, Chinese drones begin overflying the Matsu, Penghu, and Wuchiu islands. Their conduct varies, but they typically circumnavigate each major island and do not appear to be armed. The PLA releases a stream of photos taken by these drones of island installations, updating them daily on social media.

On the fourth day, in addition to the drone overflights, a single J-20 fighter aircraft flies over Kinmen, while another overflies Matsu.

Part 2

China continues conducting drone overflights and occasional fighter overflights. After a week, Beijing announces that it is extending its East China Sea ADIZ to overlap with Taiwan's ADIZ. The PRC government claims that this step is necessary in light of political air threats, including visits by foreign government officials to Taiwan masquerading as civilian flights.

The PLA also steps up its use of drones around Taiwan's main island, now venturing into waters just off the southern and eastern coasts.

Part 3

The *Global Times* publishes an editorial calling for a change in China's NFU doctrine. Including quotes from Ministry of National Defense officials, the editorial argues that the presence of nuclear-powered military ships in and around Taiwan has militarized and nuclearized these waterways. While NFU is a cornerstone of China's security, it urges that any incursion on Chinese soil or the new ADIZ by foreign military forces should be met with the severest of consequences.

Abbreviations

ADIZ	air defense identification zone
CCG	China Coast Guard
CCP	Chinese Communist Party
DPP	Democratic Progressive Party
DTRA	Defense Threat Reduction Agency
INDOPACOM	U.S. Indo-Pacific Command
KMT	Kuomintang
MND	Ministry of National Defense
NFU	No First Use
NSRD	RAND National Security Research Division
ODC	overall defense concept
PRC	People's Republic of China
PLA	People's Liberation Army
SME	subject-matter expert
TCG	Taiwan Coast Guard
TTX	tabletop exercise
UAV	unmanned aerial vehicle

References

Beauchamp-Mustafaga, Nathan, "PLA Navy Used for First Time in Naval Evacuation from Yemen Conflict," Jamestown Foundation, *China Brief*, Vol. 15, No. 7, April 3, 2015.

Blackwill, Robert D., and Philip Zelikow, "The United States, China, and Taiwan: A Strategy to Prevent War," Council on Foreign Relations, Special Report No. 90, February 2021.

Brown, Gerald C., and Ben Lewis, "Taiwan ADIZ Violations," spreadsheet, March 5, 2023.

Campbell, Brian E., "Record-Setting Incursions into Taiwan's Air Defense Identification Zone: The People's Republic of China's Psychological Operations Designed to Erode U.S. Support for Taiwan," Air University Press, *Journal of Indo-Pacific Affairs*, August 1, 2022.

Cancian, Mark F., Matthew Cancian, and Eric Heginbotham, *The First Battle of the Next War: Wargaming a Chinese Invasion of Taiwan*, Center for Strategic and International Studies International Security, January 2023.

Center for Strategic and International Studies, "Asia Maritime Transparency Initiative," webpage, undated. As of January 2, 2022:
https://amti.csis.org/

"China Paper Says Singapore Troop Carriers Should Be 'Melted Down,'" Reuters, November 28, 2016.

"China Sends Missiles Flying Over Taiwan," *The Economist*, August 4, 2022.

"Diehard 'Taiwan independence' Separatists to Be Punished According to Law: Spokesperson," Xinhua, November 5, 2021.

Doyle, Gerry, Anand Katakam, Ben Blanchard, and Marco Hernandez, "The Skies Over the South China Sea," Reuters, October 20, 2021.

Everington, Keoni, "China Commentator Says Taiwan's President Should be Assassinated," *Taiwan News*, July 15, 2022.

Faison, Seth, "China Proclaims It Designed Its Own Neutron Bomb," *New York Times*, July 15, 1999.

Gao Yang [高杨], "The Will of the People Is Irresistible" ["民意不可违 大势不可挡"], Quish [求实], June 22, 2019.

Gelpi, Christopher, Peter D. Feaver, and Jason Reifler, *Paying the Human Costs of War: American Public Opinion and Casualties in Military Conflicts*, Princeton University Press, 2009.

Goldrick, James, *Grey Zone Operations and the Maritime Domain*, Australia Strategic Policy Institute, October 1, 2018.

Green, Michael J., John Schaus, Jake Douglas, Zack Cooper, and Kathleen H. Hicks, *Countering Coercion in Maritime Asia: The Theory and Practice of Gray Zone Deterrence*, Center for Strategic and International Studies, May 9, 2017.

Heath, Timothy R., Kristen Gunness, and Tristan Finazzo, *The Return of Great Power War: Scenarios of Systemic Conflict Between the United States and China*, RAND Corporation, RR-A830-1, 2022. As of January 11, 2023:
https://www.rand.org/pubs/research_reports/RRA830-1.html

Heginbotham, Eric, Michael Nixon, Forrest E. Morgan, Jacob L. Heim, Jeff Hagen, Sheng Li, Jeffrey Engstrom, Martin C. Libicki, Paul DeLuca, David A. Shlapak, David R. Frelinger, Burgess Laird, Kyle Brady, and Lyle J. Morris, *The U.S.-China Military Scorecard: Forces, Geography, and the Evolving Balance of Power, 1996–2017*, RAND Corporation, RR-392-AF, 2015. As of January 12, 2023:
https://www.rand.org/pubs/research_reports/RR392.html

Hughes, Geraint, "War in the Grey Zone: Historical Reflections and Contemporary Implications," *Survival,* Vol. 62, No. 3, May 14, 2020.

Hunzeker, Michael A., and Alexander Lanoszka, "Real Friends Twist Arms: Taiwan and the Case for Conditionality," War on the Rocks, July 27, 2022.

Ives, Mike, and Zixu Wang, "Mostly Bluster: Why China Went Easy on Taiwan's Economy," *New York Times*, August 12, 2022.

"Japan Frees Chinese Boat Captain amid Diplomatic Row," BBC, September 10, 2010.

Jensen, Benjamin, Bonny Lin, and Carolina G. Ramos, "Shadow Risk: What Crisis Simulations Reveal About the Dangers of Deferring U.S. Responses to China's Gray Zone Campaign Against Taiwan," Center for Strategic and International Studies, February 16, 2022.

Jiyoon, Kim, John J. Lee, and Kang Chungku, "Changing Tides: THAAD and Shifting Korean Public Opinion Toward the United States and China," ASAN Institute for Policy Studies, March 20, 2017.

Lanoszka, Alexander, "Russian Hybrid Warfare and Extended Deterrence in Eastern Europe," *International Affairs*, Vol. 92, No. 1, January 2016.

Lee, Yimou, "Taiwan Defence Ministry: Website Hit by Cyber Attacks amid China Tensions," Reuters, August 3, 2022.

Lendon, Brad, "U.S. Sends Two Warships Through Taiwan Strait, in First Transit Since Pelosi Trip," CNN, August 28, 2022.

Lin, Bonny, Taylor Fravel, Cristina Garafola, Kathrin Hille, Roderick Hill, and Chris Twomey, "The Military Dimensions of the Fourth Taiwan Strait Crisis," transcript of virtual panel, Center for Strategic and International Studies, August 23, 2022.

Lin, Bonny, Cristina L. Garafola, Bruce McClintock, Jonah Blank, Jeffrey W. Hornung, Karen Schwindt, Jennifer D. P. Moroney, Paul Orner, Dennis Borrman, Sarah W. Denton, and Jason Chambers, *Competition in the Gray Zone: Countering China's Coercion Against U.S. Allies and Partners in the Indo-Pacific*, RAND Corporation, RR-A594-1, 2022. As of February 28, 2023:
https://www.rand.org/pubs/research_reports/RRA594-1.html

Lin-Greenberg, Erik, "Wargame of Drones: Remotely Piloted Aircraft and Crisis Escalation," *Journal of Conflict Resolution*, Vol. 66, No. 10, June 6, 2022.

Lin-Greenberg, Erik, Reid Pauly, and Jacquelyn Schneider, "Wargaming for International Relations Research," *European Journal of International Relations*, Vol. 28, No. 1, 2021.

Liu, Meihui, and Li-Ching Hung, "Identity Issues in Taiwan's History Curriculum," *International Journal of Educational Research*, Vol. 37, Nos. 6–7, 2002.

Liu, Nicolle, and Phelim Kine, "China's Internet Expresses Glee at Abe's Assassination," *Politico*, July 8, 2022.

Long, Drake, "Chinese Maritime Militia on the Move in Disputed Spratly Islands," Radio Free Asia, March 24, 2020.

Lubold, Gordon, "U.S. Troops Have Been Deployed in Taiwan for at Least a Year," *Wall Street Journal*, October 7, 2021.

Martin, Bradley, Kristen Gunness, Paul DeLuca, and Melissa Shostak, *Implications of a Coercive Quarantine of Taiwan by the People's Republic of China*, RAND Corporation, RR-A1279-1, 2022. As of January 11, 2023:
https://www.rand.org/pubs/research_reports/RRA1279-1.html

Martinez, Luis, "A Look at the U.S. Military's Close Calls with China, Russia in the Air and at Sea," ABC News, October 2, 2018.

Mazarr, Michael J., Joe Cheravitch, Jeffrey W. Hornung, and Stephanie Pezard, *What Deters and Why: Applying a Framework to Assess Deterrence of Gray Zone Aggression*, RAND Corporation, RR-3142-A, 2021. As of January 12, 2023:
https://www.rand.org/pubs/research_reports/RR3142.html

"Meng Wanzhou and the Two Michaels: A Timeline," *The Guardian*, September 24, 2021.

Naval History and Heritage Command, "EP-3 Collision, Crew Detainment, Release, and Homecoming," webpage, August 18, 2021. As of September 17, 2022: https://www.history.navy.mil/research/archives/Collections/ncdu-det-206/2001/ep-3-collision--crew-detainment-and-homecoming.html

Nuclear Threat Initiative, "Taiwan Overview," fact sheet, May 27, 2015.

Perla, Peter, and E. D. McGrady, "Why Wargaming Works," Naval War College Review, Vol. 64, No. 3, Summer 2011.

Perlez, Jane, and Amy Chang Chien, "Chinese Drones: The Latest Irritant Buzzing Taiwan's Defenses," *New York Times*, September 10, 2022.

Perlroth, Nicole, and David E. Sanger, "China Breached Dozens of Pipeline Companies in Past Decade, U.S. Says," *New York Times*, July 26, 2021.

Pham, Nga, "Shift as Vietnam Marks South China Sea Battle," BBC, January 15, 2014.

Phillips, Tom, and Oliver Holmes, "Activist Who Vanished in Thailand is Being Held in China, Says Wife," *The Guardian*, February 3, 2016.

Ramzy, Austin, and Chris Buckley, "'Absolutely No Mercy': Leaked Files Expose How China Organized Mass Detention of Muslims," *New York Times*, November 16, 2019.

Safeguard Defenders, "China's Hunt for Taiwanese Overseas," fact sheet, 2021.

Shan, Shelley, "Record Number of Cyber Attacks Reported," *Taipei Times*, August 5, 2022.

Shinji, Yamaguchi, Yatsuzuka Masaaki, and Momma Rira, *China's Quest for Control of the Cognitive Domain and Gray Zone Situations*, National Institute for Defense Studies, January 2023.

"Taiwan Government Faces 5 Million Cyber Attacks Daily: Official," AFP, October 11, 2021.

"The Taiwan Problem and the Project of Chinese Reunification in the New Era" ["台湾问题与新时代中国统一事业"], Qiushi [求实], August 10, 2022.

Tsui, Chin-Kuei, "China's Gray Zone Activities and Taiwan's Response," Stimson Policy Memo, Henry L. Stimson Center, December 12, 2022.

U.S. Department of Defense, "China Coast Guard Sinking of a Vietnam Fishing Vessel," press release, April 9, 2020.

U.S. Department of Defense, *China Military Power Report*, November 29, 2022.

Walker, Christopher, "What Is 'Sharp Power?'" *Journal of Democracy*, Vol. 29, No. 3, July 2018.

Walker, Christopher, "Rising to the Sharp Power Challenge," *Journal of Democracy*, Vol. 33, No. 4, October 2022.

"What Is Taiwan's Porcupine Defence Strategy?" *The Economist*, May 10, 2022.

Wuthnow, Joel, Derek Grossman, Phillip C. Saunders, Andrew Scobell, and Andrew N. D. Yang, eds., *Crossing the Strait*, National Defense University Press, 2022.

Yang, Florence Wen-Ting, "Asymmetrical Interdependence and Sanction: China's Economic Retaliation over South Korea's THAAD Deployment," *Issues and Studies*, Vol. 55, No. 4, 2019.

Yeh, Emily, *Taming Tibet*, Cornell University Press, 2013.

Zangmo, Tsultim, "The New Trends of Han Migration to TAR," *Tibet Journal*, Vol. 44, No. 1, 2019.